Get to Know Big Cats

LEOPARDS

By Bray Jacobson

Please visit our website, www.garethstevens.com. For a free color catalog of all our high-quality books, call toll free 1-800-542-2595 or fax 1-877-542-2596.

Library of Congress Cataloging-in-Publication Data
Names: Jacobson, Bray, author.
Title: Leopards / Bray Jacobson.
Description: Buffalo, New York : Gareth Stevens Publishing, [2024] | Series: Get to know big cats | Includes index. | Audience: Grades K-1
Identifiers: LCCN 2022045131 (print) | LCCN 2022045132 (ebook) | ISBN 9781538286050 (library binding) | ISBN 9781538286043 (paperback) | ISBN 9781538286067 (ebook)
Subjects: LCSH: Leopard–Juvenile literature.
Classification: LCC QL737.C23 J3372 2024 (print) | LCC QL737.C23 (ebook) | DDC 599.75/54–dc23/eng/20220929
LC record available at https://lccn.loc.gov/2022045131
LC ebook record available at https://lccn.loc.gov/2022045132

First Edition

Published in 2024 by
Gareth Stevens Publishing
2544 Clinton Street
Buffalo, NY 14224

Editor: Kristen Nelson
Designer: Leslie Taylor

Photo credits: Cover, pp. 1, 7, 24 (fur) RealityImages/Shutterstock.com; p. 5 EcoPrint/Shutterstock.com; pp. 9, 24 (spots) TigerStocks/Shutterstock.com; p. 11 Henk Bogaard/Shutterstock.com; p. 13 O'sokin/Shutterstock.com; p. 15 Tara Lambourne/Shutterstock.com; p. 17 Dr Ajay Kumar Singh/Shutterstock.com; p. 19 jeep2499/Shutterstock.com; pp. 21, 24 (cub) Ragnhild Vaa Lillehaug/Shutterstock.com; p. 23 Alta Oosthuizen/Shutterstock.com.

Printed in the United States of America

CPSIA compliance information: Batch #CSGS24: For further information contact Gareth Stevens at 1-800-542-2595.

Contents

Leopards are big cats!

They have fur.
It is yellow or white.

Their fur has dark spots.
They look like roses.

Leopards live alone.

They climb well.
They hide in trees.

Their spots help them hide.

They hunt deer
and antelope.

They are good swimmers.
They eat fish and crabs.

Babies are cubs.
They are gray at birth.

Mothers care for cubs.
Cubs stay for two years.

Words to Know

cub

fur

spots

Index